First edition, October 2018
Book design by C.B. Telemann

978-1-7328567-2-1
Seego Publishing

WATER AT THE TOP OF THE WORLD

A STORY OF LEGENDS AND LEARNING

Written by J. Siegal
Illustrations by Shannon Belock

Seego Publishing
Chicago, IL
2018

Almost everywhere you go on Earth, they have a story of a giant flood. Stories from some places say that the water rose even to the mountain tops.

In many stories, the people didn't know what caused the flood, and in other stories, the people were punished by the rising waters.

Some people say that stories themselves can move like water around the world.

hat many stories tell is that after the flood went down, there were those who lived in high places, who came down to teach the people. Some stories, from some people, say that the high ones came down from the mountains, some that they came down from the sky, from across the stars or from very near, no one is sure.

ften, people's stories said that the high ones could do wondrous things, or that they had power over the air, the water, the land, and all the creatures. Sometimes, the stories said that the high ones had knowledge from a time before the flood, and that they had saved that knowledge on high, so it would not be lost, and so it could be taught again to the people who survived the flood.

fter the flood, the people began to learn and grow. Often in the stories, the high ones taught the people ways in which to live. The stories say that the high ones had complete mastery of language, and that they spoke in wondrous, beautiful ways. Often the people did not listen.

ometimes the stories said that the high ones instructed the people to write things down. These writings were sometimes lost. Sometimes they became holy books. People often fought over the holy books, burned them, argued over them, warred over them.

Over time, the holy books passed through many people's hands. Although they were probably told not to, the people probably changed the books here and there. Some people like to pretend that this did not happen, and they like to scold people from other places for paying attention to other holy books, because they are afraid those people might have the wrong books. Sometimes, they are afraid that they might have the wrong books themselves.

This is true even though almost all the books agree on what is good: peace, love, temperance, humility, helping others, blessings, life, and respect for Creation.

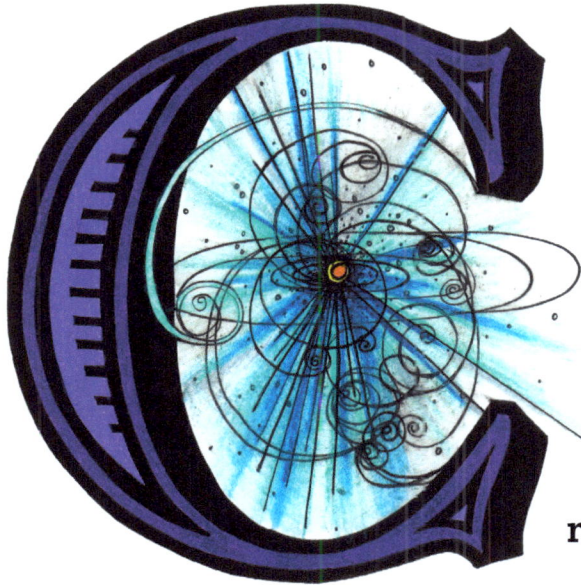

Creation.

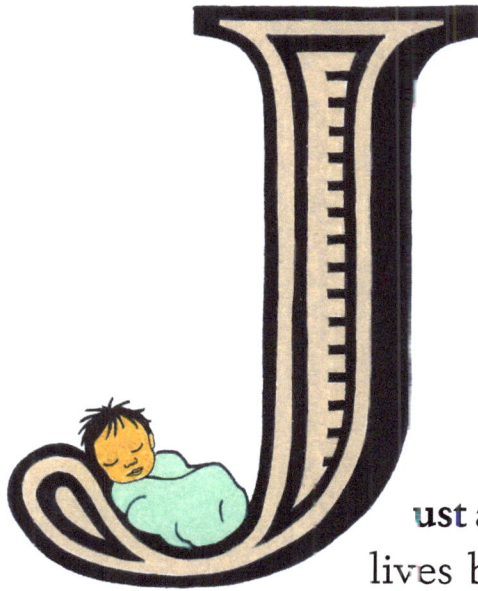

Just as a child cannot know of its parents' lives before it was born, except through stories, so the Creation cannot know of a Creator before Creation, except through stories.

Some of the holy books and some of the stories of the high ones contain tellings about Creation.

Some people think that the high one or high ones are the Creator or the Creators. Some people think that the high ones are related to or are working for the Creator. Some people think that the high ones are Creations of the Creator.

Some people think Creation just Created.

ne thing we think we have learned since the stories is that everything in Creation is moving and vibrating all the time. The very smallest things are vibrating so fast that they seem solid to us. The very big things are vibrating so slowly that they seem still to us.

There are some things so small that we can't see them, but we can see their energy. There are some things so big and far away that we can't see them, but we can see their energy. Our eyes see the vibrating light. Our ears hear the vibrating air.

What does it mean that we can see only the energy from very small or big things? It means we need help to see the very big or small things. Some people think we need help to see or feel or hear the energy of Creation. Some people think we can always see or feel or hear the energy of Creation.

Some people think the holy books can help us see or feel or hear the energy of Creation.

Quite often, the holy books contain stories about what happened to the people after they received their holy books. The holy books do not usually mention people living very far away. This is probably because the holy books passed through the hands of people, who could not see or hear or feel the other people living far away.

Now, people have help to see and hear and feel people far away. So people are learning of each other's stories.

This can be very confusing for people. This sharing of the stories makes some people very happy. It makes some people very angry.

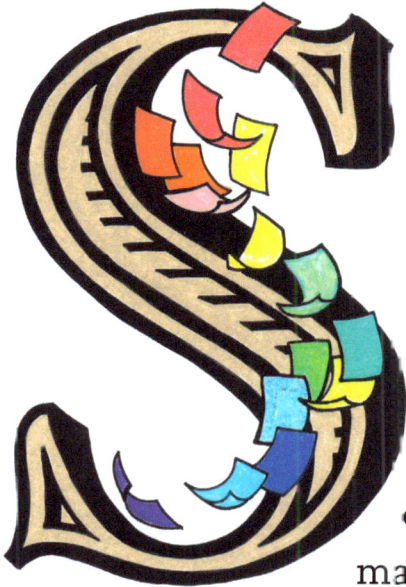

Sometimes, the sharing of the stories makes some people so angry or fearful or confused that they do things to ignore even their own holy books.

Sometimes, the sharing of the stories makes people so happy or peaceful or energetic that they search through all the stories for the best ways to live.

Sometimes, people like a holy book or stories from another place more than they like their own holy book or stories.

Sometimes, people don't want any holy books or stories at all.

They just want to see or hear or feel the energy of Creation.

APPENDIX

Some Flood Myths of the Earth

Innuit	North America	Hebrew	Middle East
Cherokee	North America	Muslim	Middle East
Navajo	North America	Masai	Africa
Hopi	North America	Pygmy	Africa
Aztec	North America	Congo	Africa
Maya	Central America	Yoruba	Africa
Carib	Central America	Mandingo	Africa
Inca	South America	Russian	Asia
Arawak	South America	Hindu	Asia
Yamana	South America	Mongolian	Asia
Greek	Europe	Tamil	Asia
Norse	Europe	Korean	Asia
Celt	Europe	Chinese	Asia
Turkish	Europe	Thai	Asia
Sumerian	Middle East	Gunwinggu	Oceania
Persian	Middle East	Maori	Oceania
Egyptian	Middle East	Fijian	Oceania
Babylonian	Middle East	Samoan	Oceania

ACKNOWLEDGMENTS

Thanks to my darling inquisitive children, E. and R. – may you light the future. Thanks to my parents and wife and family and friends for all the encouragement and love. Thanks to Alisa for some scholarly insight. Thanks to Shannon for putting heart and skill into this book.

Love, J.

For my Niece, Allison. May she always live with wisdom and learn from the times when she does not.

Love, Shannon

BIOS

J. Siegal writes fiction, nonfiction, music, code, and recipes. He plays barrelhouse piano and pulls occasional stints in world music projects. His writing has appeared in Michigan Quarterly Review and Skeptic Magazine. Currently, he is at work on his first novel. He lives with his wife and two children near Chicago, IL.

Shannon Belock is a professional free-lance artist from Tucson, Arizona who received her BFA at The School of the Art Institute of Chicago (SAIC). Though her favorite medium is painting murals, she also enjoys print making, book making, illustrating, animating, fibers and collage. Her work is inspired by health, humor & happiness. You can find more of her creations online by searching for her art pseudonym, "Heart of an Astronaut".

For additional copies, art prints, gifts, and other info about Water at the Top of the World, visit: **https://wateratthetopoftheworld.com/**

www.ingramcontent.com/pod-product-compliance
Lightning Source LLC
Chambersburg PA
CBHW040024050426
42452CB00003B/130